Animals of the Sea

BY JOY BREWSTER

Table of Contents

1. Oceans As Home 2
2. Go Fish! 8
3. Meet the Mammals 12
4. Incredible Invertebrates 16
5. Saltwater Reptiles 20
6. The Amazing Sea Animal Hall of Fame 24
Solve This! Answers 30
Glossary 31
Index 32

CHAPTER 1

Oceans As Home

Did you know that oceans cover almost three-fourths of our planet? These oceans vary from tropical waters to arctic seas. Yet each is home to an abundance of living things.

In every part of the ocean live unique marine animals, from the tiniest **plankton** to the blue whale, the largest animal on Earth. Many of these species are colorful and strange, like **fish** that flash in the darkness or mysterious giant squid that no one has ever seen alive.

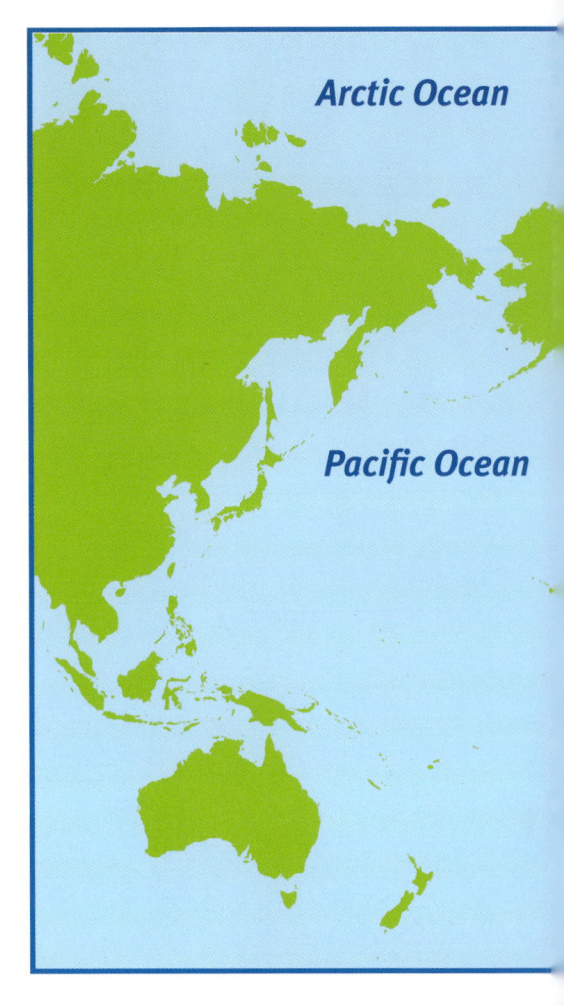

Arctic Ocean

Pacific Ocean

blue whale

sea urchin

Arctic Ocean

Atlantic Ocean

Pacific Ocean

Indian Ocean

giant squid

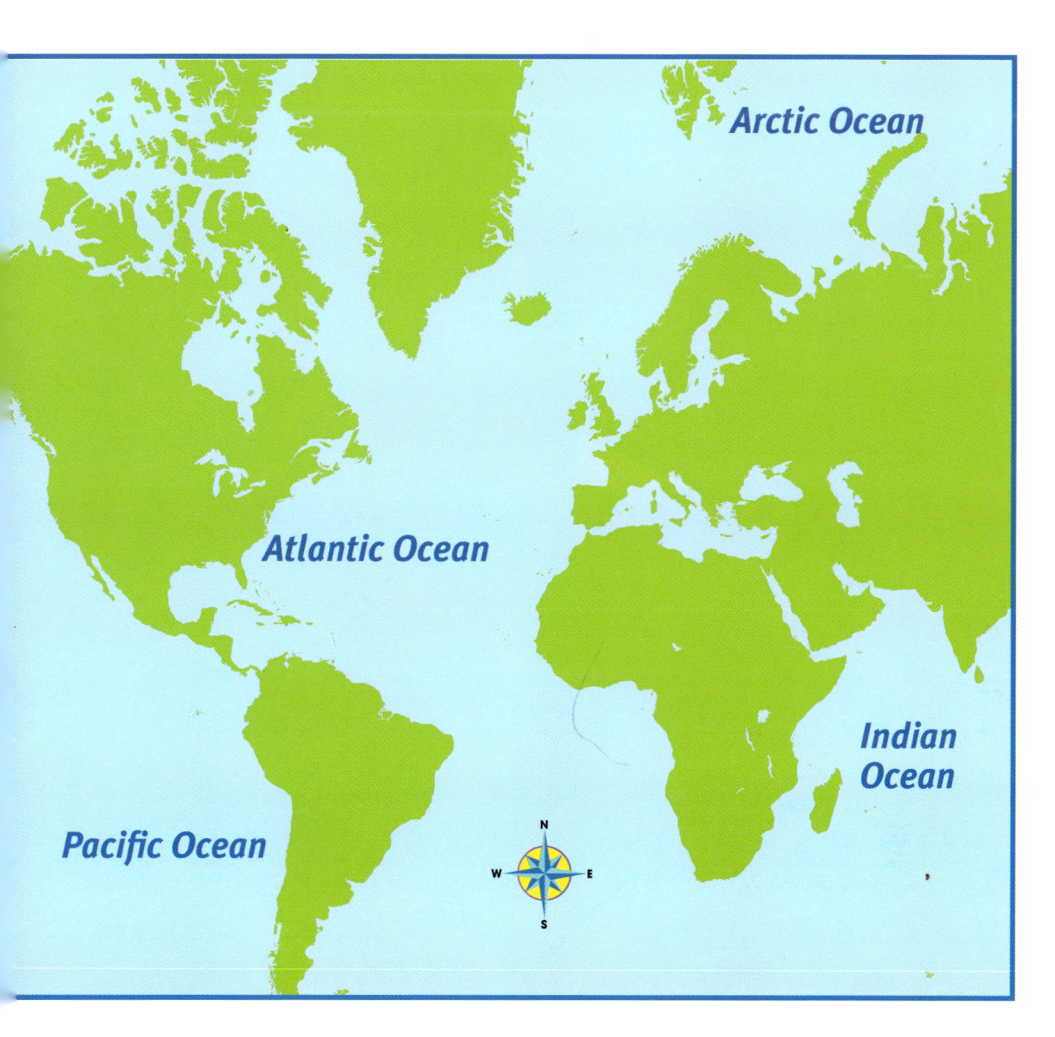

THE OCEANS' TINIEST CREATURES

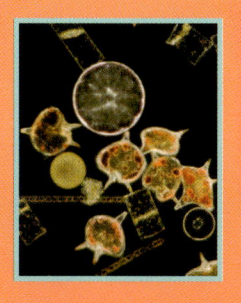

"Plankton" is a general word for the microscopic plants and animals that drift in the ocean. These tiny plants and animals are an important link in the food chain. Almost all ocean creatures depend on plankton to survive.

CHAPTER 1

↑ **These bottle-nosed dolphins frolic in the sea.**

Imagine that you are diving down into the ocean. As you travel more than six miles from the surface to the ocean floor, you pass through different levels. You notice changes in the light, temperature, and pressure. As you go deeper, the water gets darker and colder and the pressure increases. Each level presents a unique habitat where different kinds of animals live.

Scientists describe these ocean levels as "zones." The three main zones are the **sunlight zone**, the **twilight zone**, and the **midnight zone**.

Most marine animals live in the sunlight zone because they can eat the plants that grow there. This is the only zone where enough sunlight penetrates the water to support plant life.

OCEANS AS HOME

Small animals live in the dark and murky twilight zone. Creatures there must prey on other animals or eat the dead animals and plants that fall from above.

Some very unusual creatures live in the midnight zone where it is pitch-black, near freezing, and the water pressure is tremendous. Down here are hot water vents—cracks in the ocean floor that shoot out incredibly hot water filled with minerals. Living near these vents are large, wormlike animals called tube worms. Tube worms get their nutrients from bacteria living inside them. The bacteria, in turn, thrive on the heat and minerals.

The bottle-nosed dolphin is one animal that lives in the sunlight zone. It is an extremely intelligent and friendly mammal. The dolphin relies mostly on its strong sense of hearing to find its way around. It can leap thirty feet in the air!

↑ tube worms

CHAPTER 1

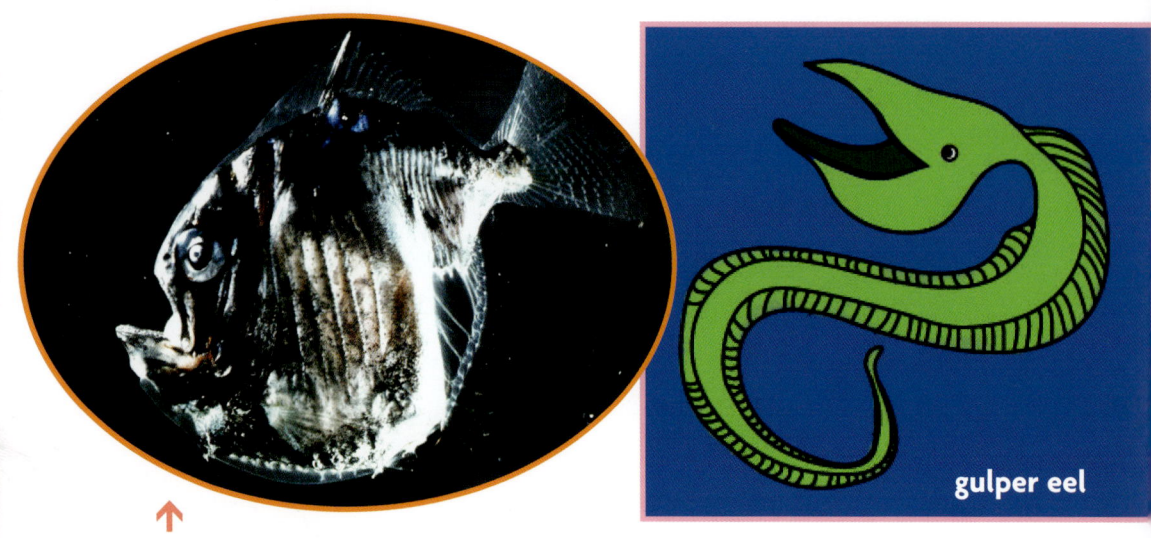

The hatchetfish uses camouflage to protect itself.

gulper eel

SOLVE THIS!

1 If one inch equals 2.54 centimeters, about how many centimeters long is a hatchetfish?

A gulper eel enjoys a hearty meal on February 15 but has to wait another two weeks before it can find another. If it's not leap year, what will the date be?

The silver hatchetfish is an animal that lives in the twilight zone. It has large eyes that allow it to see in the murky darkness. Because it is only about three inches long and its body gives off a blue light, this fish is nearly invisible.

The gulper eel is an animal that lives in the midnight zone. It has an enormous mouth and an expandable stomach. It can swallow fish and other creatures bigger than itself. Because not much food is available in the midnight zone, creatures that live there don't eat very often. When they do, they need to be able to eat enough to last them a long time.

OCEANS AS HOME

SOLVE THIS!

2. Which is the largest ocean zone?

SUNLIGHT ZONE

0 to 650 feet
(0 to 198 meters)

TWILIGHT ZONE

650 to 3,300 feet
(198 to 1,006 meters)

MIDNIGHT ZONE

3,300 to 36,198 feet,
(1,006 to 11,033 meters)
lowest point on
ocean bottom

7

CHAPTER 2

Go Fish!

More than 13,300 kinds of fish live in the ocean. They come in all shapes, sizes, and colors—and some of them have amazing talents! Here are some interesting facts about fish:

- All fish breathe underwater through gills.
- All fish use fins to swim.
- All fish are **cold-blooded**, which means that their temperatures change with the temperature of the surrounding water.
- Most fish are covered with scales.
- Most fish have bony skeletons.
- Most fish lay eggs.

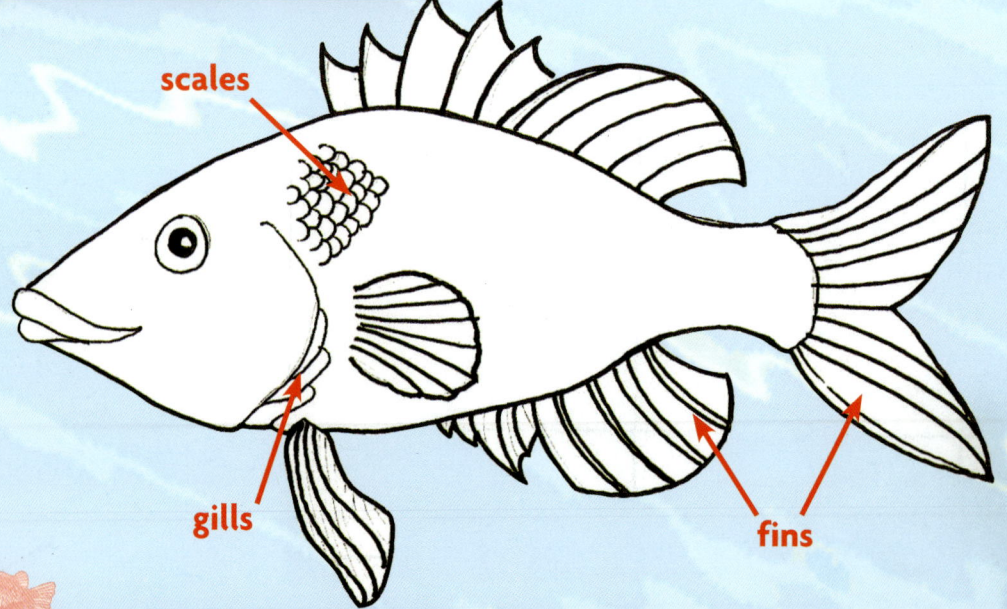

egg

scales

gills

fins

Fish have many **predators**, including whales, seals, and other fish, so they need to be able to protect themselves. Different kinds of fish protect themselves in different ways.

It's no wonder where the colorful lionfish gets its name! Lionfish use their deadly, poisonous spines to keep away predators.

The puffer also has spines. It can blow itself up like a balloon, making its sharp spines stick out. This helps keep away its predators, such as sharks!

Lantern fish sparkle with tiny lights. This **bioluminescence** (by-oh-loo-mih-NEH-sens) helps them find food or confuse prey. Bioluminescence is most common in the creatures that live in the deepest, darkest parts of oceans.

school of fish

lionfish

puffer

lantern fish

CHAPTER 2

Fish Close-Up: Sharks

NOTEWORTHY NEWS ABOUT SHARKS

Did you know that sharks are part of the fish family? Like other fish, they're cold-blooded, have fins, and breathe through gills. Here are a few more things you may be surprised to learn about sharks:

- Sharks have been swimming in Earth's oceans from before the age of the dinosaurs.

- Most sharks do not lay eggs. They give birth to live babies.

- Sharks often prey in dark, murky waters so they have to rely on their powerful senses to find food. They can hear animals up to 3,000 feet (914 meters) away. They can detect the direction of the faintest odors. Their eyesight is about seven times better than that of humans.

- Sharks don't have scales like other fish. A shark's skin is covered with tiny thorns, so it is rough and prickly.

- Sharks don't have a swim bladder, which is a special organ that helps fish stay afloat. That means that if sharks stop swimming, they'll sink!

- Sharks are always losing their teeth. But whenever they lose a tooth, a new one takes its place. Some sharks go through 30,000 teeth in a lifetime.

white shark

hammerhead shark

mako shark

GO FISH!

whale shark

SOLVE THIS!

3 If a shark lives for 60 years and loses 30,000 teeth, what is the average number of teeth that the shark loses each year?

Some sharks can eat up to 10% of their body weight in a day. At this rate, how much would a 250-pound (113-kilogram) shark eat in a week?

What is the difference in size between the largest and the smallest shark in the ocean?

If a whale shark is 50 feet (15 meters) long and weighs 55,000 pounds (24,945 kilograms) and its weight is evenly distributed, how much would one foot of whale shark weigh?

There are over 250 different kinds of sharks, from the gigantic whale shark, which can grow to be 50 feet (15 meters) long, to the smallest shark in the ocean, which is only about 6 inches (15 centimeters) long.

The whale shark is the biggest shark in the ocean. It can weigh up to 55,000 pounds (24,945 kilograms) — that's as much as a bulldozer! Given its size, it's surprising to learn that the whale shark is very gentle and feeds mostly on tiny plankton.

CHAPTER 3

Meet the Mammals

The Weddell seal can dive to almost 2,000 feet (610 meters) and hold its breath for almost an hour.

Some marine animals, such as whales, dolphins, and seals, are not fish. They may swim and hunt in the ocean like fish, but they're **mammals**. Mammals include dogs, tigers, and humans. Here are some interesting facts about mammals:

- Mammals have lungs and breathe air, which means that they can't breathe underwater.
- Mammals are **warm-blooded**, which means that their body temperature stays the same even if the surrounding temperature changes.
- Mammals give birth to live young and nurse their babies with milk.
- Mammals have hair on their bodies.

The northern fur seal swims a round trip of 4,000 miles (6,437 kilometers) each year when it migrates.

← The Pacific walrus can weigh as much as 2,000 pounds (907 kilograms). It is usually timid and shy, but it can be easily provoked.

SOLVE THIS!

4 If a northern fur seal migrates 20 times during its life, how many miles will it have covered in all?

If you saw 5 male walruses and each one had tusks that were 2.5 feet (0.76 meters) long, how many feet (meters) of tusk would there be in all? Remember that walruses have two tusks each!

There are more than thirty kinds of seals and sea lions, all of which are mammals. Most seals are graceful in the water but have to wriggle along on their bellies when ashore. Sea lions, on the other hand, use their front flippers to waddle quickly on land. Male walruses have tusks that can grow to three feet long! The male with the longest tusks usually becomes the leader of the herd. Female walruses give birth to one calf every other year.

A thick layer of fat called blubber lies just under the skin of seals, sea lions, and walruses. Blubber reduces the amount of body heat lost to the surroundings. It also contributes to the streamlined shape of these marine mammals.

CHAPTER 3

Mammal Close-Up: Whales

THE GIANTS OF THE SEA

Think you don't have anything in common with a whale? Think again! They're mammals, too.

There are two kinds of whales: baleen whales and toothed whales.

Whales that do not have teeth are called baleen whales. They filter (or strain) their food out of the water. They eat small fish and plankton. Not all baleen whales are big, but some are as long as 110 feet.

Toothed whales prey on big sea animals so they are usually good swimmers and hunters. Different kinds of toothed whales have different kinds of teeth. It all depends on what kind of food they eat. The bigger the prey, the bigger the teeth!

Male whales are called bulls, female whales are called cows, and young whales are called calves.

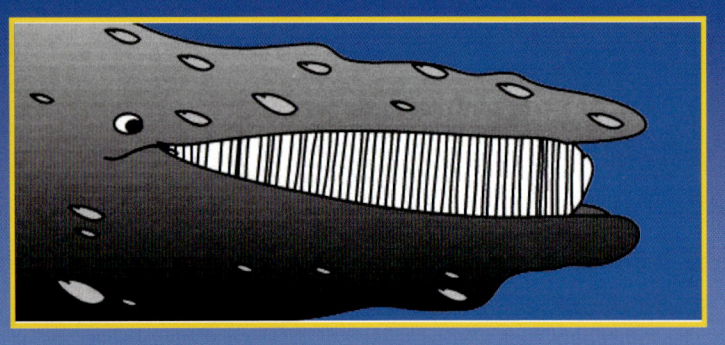

← A baleen whale swims through a plankton patch with its mouth open, forcing the water through the baleen plates and trapping the plankton on the bristles that line the inner surface.

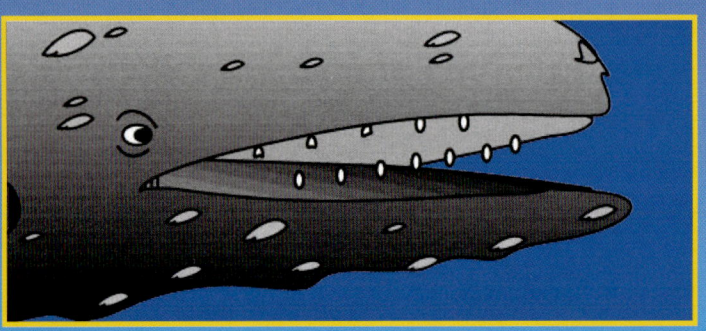

← Toothed whales that eat little fish tend to have smaller teeth. What sized teeth do you think a whale that eats large prey has?

14

MEET THE MAMMALS

HOW BIG ARE THESE COMMON WHALES?

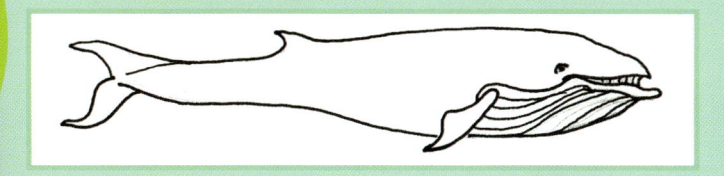

↑ blue whale:
110 feet
(34 meters)

← sperm whale:
69 feet
(21 meters)

← bowhead whale:
65 feet
(20 meters)

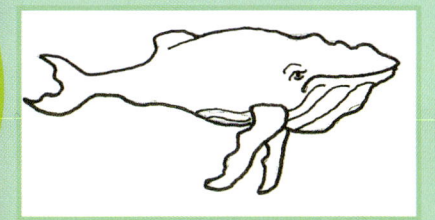

← humpback whale:
62 feet
(19 meters)

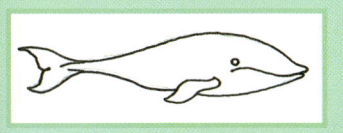

← gray whale:
50 feet
(15 meters)

← killer whale:
32 feet
(8 meters)

SOLVE THIS!

5 Which whale is about half the size of the blue whale?

Which two whales are closest in size?

Whales swim in groups of about 30, called pods. If you saw 20 pods, about how many whales would there be in all?

CHAPTER 4

Incredible Invertebrates

Backbones support our bodies. Without a backbone, we couldn't stand up straight or walk. Many creatures in the ocean, such as fish and sea mammals, also need backbones for support and movement. But some animals, called **invertebrates**, don't have backbones.

Many of the strange creatures of the sea are invertebrates. Like fish and sea mammals, invertebrates come in many shapes, sizes, and colors.

Jellyfish and sea anemones use long **tentacles** to sting their prey. Jellyfish can swim, but mostly they just drift along with the currents. Sea anemones look like colorful flowers. They attach themselves to rocks.

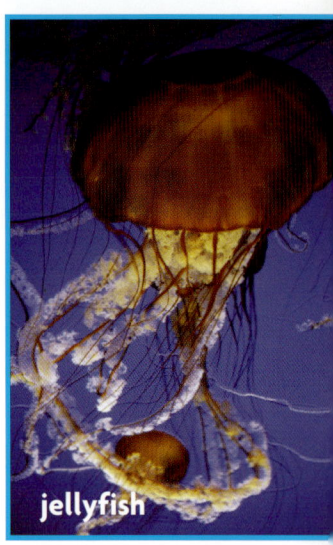

jellyfish

sea anemone

SOLVE THIS!

6. Imagine you are an underwater explorer. During one deep-sea dive, you come across three kinds of invertebrates—squid, starfish, and crabs. If you see 12 tentacles, 18 claws, and 63 arms, how many of each animal have you encountered? (Don't forget, squid have arms, too!)

The octopus and squid move through the ocean by means of **jet propulsion**—sucking water into their bodies and then forcing it out through a narrow hole. Both the octopus and the squid have eight arms, but a squid also has two long tentacles used to catch prey.

Lobsters and crabs can walk along the ocean floor. Their bodies are soft, but they have hard shells to protect them from predators. Most crabs have ten legs—two that end in big claws for capturing prey.

Starfish and sea urchins have spiny skin and tube-shaped feet that allow them to walk along the sea floor. These unusual creatures have mouths on their underside. A typical starfish has five arms. Sea urchins are covered with sharp spikes that keep their enemies away.

octopus

squid

lobster

crab

starfish

sea urchin

CHAPTER 4

Invertebrate Close-Up: Giant Squid

MYSTERIOUS MARINE MONSTERS

The giant squid is one of the biggest mysteries of the marine world. These enormous creatures live in such deep, dark ocean water that no one has ever seen a live one. Scientists have been able to piece together some information from dead ones that have washed ashore or been caught in nets.

Like smaller squid, the giant squid is an invertebrate—the largest invertebrate on Earth! It can grow up to 60 feet (18 meters) long and weigh more than 1,000 pounds (454 kilograms). It feeds on fish, smaller squid, and even some whales.

Despite its awesome size, the giant squid still has its predators, such as large sperm whales. One way it defends itself is by shooting ink into the water. The ink distracts the predator while the giant squid quickly swims away.

↑ Scientists study this dead squid as a way to learn more about these marine mysteries.

SOLVE THIS!

7 The giant squid is one of the fastest animals in the ocean. It can move both forward and backward at a speed of 35 miles (56 kilometers) per hour.

How many hours would it take a giant squid to travel 175 miles (282 kilometers)?

INCREDIBLE INVERTEBRATES

GET A CLOSER LOOK AT THE GIANT SQUID

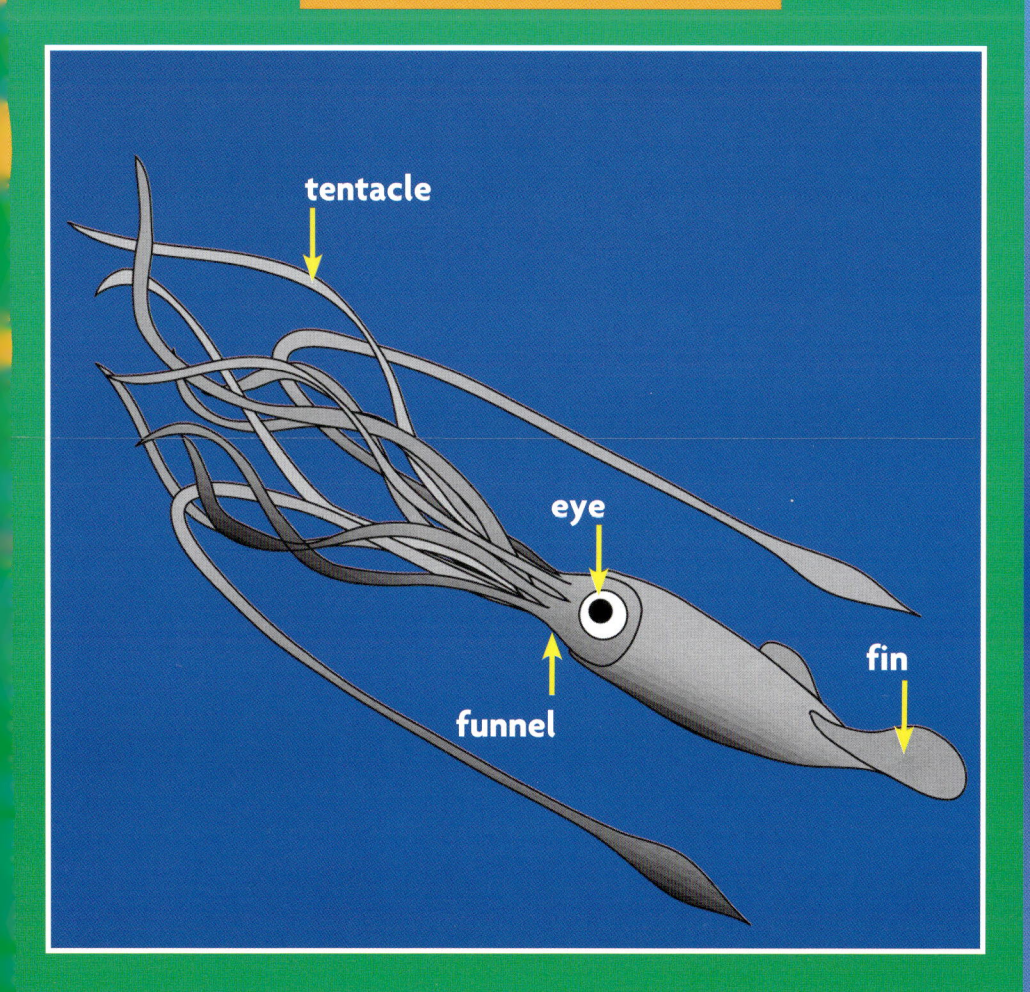

tentacle

eye

funnel

fin

Tentacles: In addition to eight tentacles used to catch and eat prey, the giant squid has two long arms with suckers on the ends.

Eyes: Its eyes are about the size of a human head, the largest eyes of any creature on Earth.

Funnel: The giant squid moves by forcing a jet of water out of a funnel, propelling itself through the water.

Fin: A fin helps the giant squid balance and turn.

CHAPTER 5

Saltwater Reptiles

Not all marine animals can be classified as fish, mammals, or invertebrates. Some sea animals, such as sea turtles and saltwater crocodiles, are **reptiles**. Reptiles, a group of animals that includes snakes, lizards, turtles, alligators, and crocodiles, share these characteristics:

- Most reptiles are covered with dry scales.
- Reptiles breathe through their lungs.
- Reptiles eat both plants and animals.
- Most reptiles reproduce by laying eggs.
- Most reptiles are cold-blooded.

Turtles lay eggs, and sea turtles are no exception. Although sea turtles live in the ocean, adult females will come ashore at night to "nest," burying up to one hundred eggs in the beach sand. Many sea turtles return to their birth place to lay eggs.

A female may nest several times in one season, but will usually wait another two to three years before nesting again.

↑
sea turtle eggs and babies on the beach

It takes about fifty to seventy days for the baby turtle to mature and break out of its egg. Then at night, it emerges from the nest and heads for the sea with the other hatchlings. At night, there's less chance of drying out from the sun and the hatchlings are safer from predators.

SOLVE THIS!

8) If there are 2,000 eggs at a nesting site and only 200 hatchlings reach the water, what is the survival rate?

CHAPTER 5

Reptile Close-Up: Leatherback Turtle
THE WORLD'S LARGEST TURTLE

The leatherback sea turtle is more than twice as large as any other species of sea turtle. Leatherbacks are also bigger than any turtle on land, making them the largest turtles in the world. Leatherbacks grow to about eight feet long and weigh almost 2,000 pounds (907 kilograms)!

There are seven species of sea turtles in the world. Look at the chart on page 23 to see how the leatherback compares to the others in size.

In addition to its enormous size, the leatherback has other distinct characteristics:

- Its shell is like hard rubber, without the scales found on other sea turtles.
- Leatherbacks have weak jaws and are adapted to their diet of jellyfish and other soft-bodied creatures.
- Leatherbacks have the largest range of all sea turtles and can be found as far north as Alaska.
- Leatherbacks have the longest migrations of all turtles, sometimes traveling more than 3,000 miles (4,828 kilometers) to their nesting sites.

ENDANGERED!

All species of sea turtles are endangered, including the leatherback. There are many reasons for their declining numbers:

- When baby sea turtles head for the ocean, they instinctively move toward the light—usually created by moonlight on the sea. However, bright lights from buildings and hotels can often confuse sea turtles so that they walk in the wrong direction, never making it to the sea.
- Adult sea turtles die when they eat plastic garbage that's floating in the ocean. They mistake the plastic for jellyfish.
- Many sea turtles get caught in fishing nets and drown.

HOW BIG ARE THESE TURTLES?

leatherback sea turtle

→ leatherback
(96 inches/244 centimeters)

← green sea turtle
(40 inches/102 centimeters)

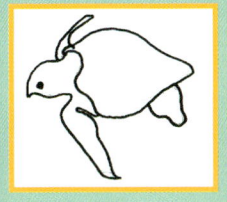

← Australian flatback turtle
(39 inches/99 centimeters)

← loggerhead turtle
(38 inches/97 centimeters)

← hawksbill turtle
(36 inches/91 centimeters)

← olive ridley turtle
(27 inches/69 centimeters)

SOLVE THIS!

9. Which turtle is about forty percent as long as the leatherback?

Which turtle is three feet (0.91 meter) long?

CHAPTER 6

The Amazing Sea Animal
HALL OF FAME

LARGEST ANIMAL

The blue whale is the largest animal ever to live on Earth—including any of the dinosaurs! The blue whale can grow up to 110 feet long and weigh up to 380,000 pounds (172,365 kilograms).

blue whale

SMALLEST FISH

The dwarf pygmy goby is about ⅓ inch long. This transparent, nearly-invisible fish lives in fresh water most of the time but goes to salt water to hatch its eggs.

SOLVE THIS!

10 If an elephant weighs about 10,000 pounds (4,536 kilograms), how many elephants would it take to equal the weight of one blue whale?

How many dwarf pygmy gobies would it take to equal the length of one blue whale?

LARGEST GROWTH

An adult ocean sunfish can grow to 500 times its birth size and can weigh as much as 3,000 pounds (1,361 kilograms).

sunfish

SLOWEST SWIMMING FISH

Seahorses are the slowest fish of the ocean, moving at about $1/10$ mile per hour ($1/16$ kilometer per hour). These unusual fish often wrap their tails around plants to keep from being swept along by a fast current.

seahorse

SOLVE THIS!

11 If you weighed 8 pounds (4 kilograms) when you were born and grew at the same rate as the ocean sunfish, how much would you weigh as an adult?

If one mile is 5,280 feet, about how long would it take a seahorse to travel one foot?

CHAPTER 6

FASTEST SWIMMING FISH

Fish		Speed
sailfish		68 miles per hour (109 kilometers per hour)
marlin		50 miles per hour (80 kilometers per hour)
bluefin tuna	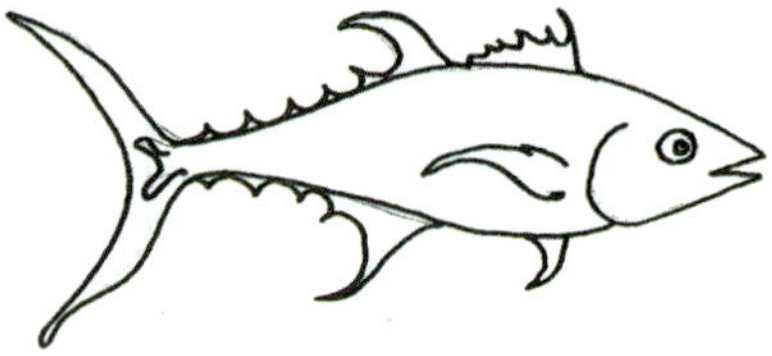	46 miles per hour (74 kilometers per hour)
yellowfin tuna		44 miles per hour (71 kilometers per hour)
blue shark		43 miles per hour (69 kilometers per hour)

THE AMAZING SEA ANIMAL HALL OF FAME

FANTASTIC FLYER

Flying fish have long fins that help them remain airborne for as long as 20 seconds.

In the water, they reach a top speed of only about 23 miles per hour (37 kilometers per hour). Once airborne, however, they can fly as fast as 35 miles per hour (56 kilometers per hour)!

flying fish

SOLVE THIS!

12 If one mile equals 1.6 kilometers, which fish swims 80 kilometers per hour?

About how long would it take a flying fish to swim 58 miles (93 kilometers)? If it could fly the whole time, how long would it take a flying fish to fly the same distance?

sperm whale

DEEPEST DIVER

The sperm whale can dive deeper than any other mammal. It lives at the surface, but can dive more than a mile (1.6 kilometers) underwater.

27

CHAPTER 6

HORRIBLE HUNTER

Sand tiger sharks hunt even before they are born! There are usually from ten to fifteen embryo sharks in the mother's womb. As these embryos develop, they eat each other until there are only two remaining!

sand tiger shark

OUCH…TERRIFYING TENTACLES

The Australian sea wasp has the most painful sting of any animal. The tentacles on this jellyfish can reach 120 feet long (37 meters)!

sea wasp

SOLVE THIS!

13 If there are 10 sand tiger embryo sharks and only 2 survive, what is the survival rate?

Which animal lives about three times longer than the sea lion?

If the average lifespan of a woman living in the United States is 80 years, by how many years do clams outlive women?

THE AMAZING SEA ANIMAL HALL OF FAME

AVERAGE LIFESPAN

Animal		Lifespan
quahog (clam)		200 years
killer whale		90 years
sea anemone		80 years
whale shark		60 years
blue whale		45 years
sea lion		30 years
starfish		15 years

Solve This! Answers

1. **Page 6**
7.62 centimeters;
March 1

2. **Page 7**
the midnight zone

3. **Page 11**
500 teeth;
175 pounds (79 kilograms);
49 feet 6 inches (594 inches);
1,100 pounds

4. **Page 13**
80,000 miles (128,748 kilometers);
25 feet (7.62 meters)

5. **Page 15**
gray whale;
bowhead whale and humpback whale;
600 whales

6. **Page 16**
3 starfish, 6 squid, and 9 crabs

7. **Page 18**
5 hours

8. **Page 21**
10%

9. **Page 23**
the loggerhead turtle;
the hawksbill turtle

10. **Page 24**
38 elephants;
3,960 dwarf pygmy gobies

11. **Page 25**
4,000 pounds (1,814 kilograms);
about 6.8 seconds

12. **Page 27**
the marlin; about 2.5 hours;
a little over an hour and a half

13. **Page 28**
20%;
the killer whale;
120 years

Glossary

bioluminescence (by-oh-loo-mih-NEH-sens) the light given off by living things

cold-blooded (KOLD-BLUH-ded) having a body temperature that changes with the temperature of the environment

fish (FISH) a marine animal that breathes using gills, lays eggs, and is usually covered with scales

invertebrates (in-VER-teh-bruts) animals without backbones

jet propulsion (JET pruh-PUL-shun) movement of an object that is caused by pushing liquid in the opposite direction

mammals (MA-mulz) warm-blooded animals that have hair on their bodies, give birth to live young, feed their young milk, and breathe with lungs

midnight zone (MID-nite ZONE) the deepest layer of the ocean

plankton (PLANK-tun) microscopic plants and animals that float in the ocean

predators (PREH-duh-terz) animals that feed on other animals

reptiles (REP-tilez) cold-blooded animals that lay eggs and have backbones

sunlight zone (SUN-lite ZONE) the top layer of the ocean, where most marine animals live

tentacles (TEN-tuh-kulz) thin, flexible parts of octopi and jellyfish

twilight zone (TWY-lite ZONE) the middle layer of the ocean

warm-blooded (TWY-lite ZONE) having a body temperature that always stays about the same

Index

Australian sea wasp, 28
baleen whale, 14
bioluminescence, 9
cold-blooded, 8
crab, 17
dolphin, 4–5, 12
dwarf pygmy goby, 24
eggs, 8, 10, 20, 21, 24
fish, 2, 8–11, 14, 16, 18, 20, 24–29
flying fish, 27
gills, 8, 10
gulper eel, 6
hatchetfish, 6
invertebrates, 16–20
jellyfish, 16, 22, 28
jet propulsion, 17
lantern fish, 9
lionfish, 9
lobster, 17
mammal, 5, 12–15, 16, 20, 27
midnight zone, 4–7
minerals, 5
octopus, 17

plankton, 2, 3, 11, 14
predator, 9
pressure, 4–5
puffer, 9
reptiles, 20–23
scales, 8, 10, 20, 22
sea anemone, 16, 29
sea horse, 25
sea lions, 13, 29
sea urchin, 17
seal, 9, 12, 13
shark, 9–11, 26, 28–29
squid, 2, 17, 18–19
starfish, 17
sunfish, 25
sunlight zone, 4–5, 7
tentacles, 16, 17, 19, 28
toothed whale, 14
turtle, 20–23
twilight zone, 4, 6–7
walrus, 13
warm-blooded, 12
whale, 12, 14–15, 18, 24, 27, 29